Daysy Cristina dos Santos Lorena
Romulo da C. da Silva
Carlos R. Franzini F.º

Analysis of the collaboration of waste generators in selective collection

Daysy Cristina dos Santos Lorena
Romulo da C. da Silva
Carlos R. Franzini F.º

Analysis of the collaboration of waste generators in selective collection

An analytical study into the collaboration and awareness of waste generators in the selective collection process

ScienciaScripts

Imprint
Any brand names and product names mentioned in this book are subject to trademark, brand or patent protection and are trademarks or registered trademarks of their respective holders. The use of brand names, product names, common names, trade names, product descriptions etc. even without a particular marking in this work is in no way to be construed to mean that such names may be regarded as unrestricted in respect of trademark and brand protection legislation and could thus be used by anyone.

Cover image: www.ingimage.com

This book is a translation from the original published under ISBN 978-620-2-19537-9.

Publisher:
Sciencia Scripts
is a trademark of
Dodo Books Indian Ocean Ltd. and OmniScriptum S.R.L publishing group

120 High Road, East Finchley, London, N2 9ED, United Kingdom
Str. Armeneasca 28/1, office 1, Chisinau MD-2012, Republic of Moldova, Europe
Printed at: see last page
ISBN: 978-620-7-30616-9

SUMMARY

SUMMARY

This thesis is based on the study of the behavior of the population with regard to the disposal of waste produced by them. As a contribution to reducing this waste, this research aims to analyze the level of collaboration and awareness of the population, focusing on the importance of solid waste collection. It is intended that this study can show the reality of people's collaboration and knowledge of selective waste collection, so that it can serve as a basis for future projects to raise awareness of selective solid waste collection to benefit environmental sustainability, sensitizing them to change their habits, undertaking strategies and actions that promote changes in attitude, for the creation of educational spaces that strengthen the moral, intellectual and civic condition of the population and, therefore, provide a solution to the problems of pollution in the environment.

Keywords : collaboration, awareness, selective collection, solid waste.

1 INTRODUCTION

In the Middle Ages, handicrafts were the predominant form of production, with reasonable profits and low-scale production. These disadvantages of craft work, together with the bourgeoisie's desire for higher profits, led to this type of manual labor being replaced by manufacturing, with England pioneering this change.

In the 18th century, the Industrial Revolution began in England, accompanied by technological advances in machinery and transportation, generating accelerated, large-scale production (IGLESIAS, 1982). This model of large-scale production made products cheaper and increased profits. Since then, production has become easier, with machines working faster than men, as if nothing were impossible. The effects of this fast, carefree and large-scale production soon generated some discomfort, especially among workers who were dissatisfied with the working day and the unemployed who were replaced by machines. At the same time, nature was ignored, seen only as a cheap supplier of endless raw materials (IGLESIAS, 1982).

In the centuries before the Industrial Revolution, nature was able to neutralize the harmful effects that overexploitation and man-made waste had on nature. However, since the 1950s, this scenario has changed due to the rapid growth of the world population (RICHARD KOLLODGE, 2011). As a result, consumption grew exponentially and industries began to consume natural resources excessively, generating negative impacts in terms of resource capture and final product disposal.

At this time, Brazil was keeping pace with world consumption, adopting a high-standard consumption model. According to data from the Brazilian Institute of Geography and Statistics (IBGE), in 1989 the country produced 241,614 tons of garbage every day, 75% of which went to the garbage dump, 0.7% to flooded dumps, with only 23.3% receiving adequate treatment, and the rest having an unknown destination (IBGE, 1989).

In the 2000s there was an improvement in the final disposal of Brazilian waste. 47% of the waste collected daily was sent to landfills, 22.3% was sent to controlled landfills, where it received appropriate treatment, and the remaining 30.5% was sent to garbage dumps, but the majority of Brazilian municipalities still used garbage dumps as the final destination for their waste (IBGE, apud VIEIRA, 2008).

In Brazil, the creation of the National Solid Waste Policy (PNRS) in 2010 (Law 12.305/2010) aimed to boost recycling and the reuse of solid waste, with the proposal of sustainable consumption that provides for the reduction of waste generation in the country. This stimulus is proven by a survey by Ciclosoft (2016), which found an increase in the number of municipalities with selective collection in the country. These totaled 443 in 2010, 927 in 2014, and 1055 in 2016, resulting in 18% of all Brazilian municipalities, and of these 1055 municipalities, 54% have cooperatives as the institution carrying out this collection.

These municipal selective collection programs are still very much concentrated in the Southeast and South regions of the country. 81% of the municipalities that carry out this service are located in these regions, totaling 31 million Brazilians served by selective collection, or 15% of the population. This reality is demonstrated by research commissioned by the Business Commitment to Recycling (CEMPRE, 2016).

Currently, the three existing selective collection models are: those carried out by the municipality itself; companies contracted to carry out the collection; and waste pickers' cooperatives, the latter being maintained or supported by companies through sorting sheds, trucks, training, machinery and help with water and electricity costs (CEMPRE, 2016).

Seeing selective waste collection as a chain, there are agents who must be active in their roles so that this chain doesn't break and its objective is achieved. This agent is the waste generator, who consumes before generating. This high standard of consumption also influences sustainability.

According to Portilho (2005), until the 1970s the environmental crisis was attributed to population growth. It was only in 1973, at the Stockholm conference, that the production models of industrialized nations were seen as the root cause of the environmental crisis. Then, at the United Nations Conference on Environment and

Development, also known as Rio92, the unequal contribution of different lifestyles and consumption to environmental problems made the debate on the relationship between consumption and the environment one of the main issues in the search for sustainability.

Consumption is understood as a field of political and symbolic struggles guided by culture, and it is possible to understand that consumption is linked to political choices, preference for certain goods, lifestyle, the idea of nature and the economy of society, in other words, consumption is not an individual preference and the consumer is not manipulated by a company's marketing. On the contrary, the consumer is active in the production and reproduction of social values, the construction and strengthening of identity, distinction, cultural hostility and cultural distress (PORTILHO, 2005).

In view of these initial considerations, selective collection is seen as a way of mitigating the problem of tons of recklessly discarded waste. Since the population is the agent of excessive consumption, which consequently generates waste, this work focuses on raising awareness and collaboration among waste generators. Knowing that some of the ways to reduce this culture of consumption and careless disposal are changes in awareness (knowing the impact of selective collection on their daily lives and on sustainability) and collaboration (how active the generator is when it comes to exercising its role in the selective collection chain), this work proposes to investigate, as a central point of its development, the level of awareness and collaboration of individual waste generators in the selective collection process.

1.1 Objectives

1.1.1. General Objective

The general objective of this work is to analyze the level of awareness and collaboration of individuals who generate waste in the selective collection process.

1.1.2. Specific Objectives

Below are the specific objectives considered for this work.

- A bibliographical survey on the topics of selective collection, collaboration, awareness and solid waste, as well as a documentary survey of organizations and associations in the sector;
- Identify in the literature the aspects and factors that determine the awareness and collaboration of individual waste generators in the selective collection process;
- Propose questions and assertions to analyze the level of awareness and collaboration of individuals who generate waste in the selective collection process;
- To verify the awareness and collaboration of individuals who generate waste by applying an exploratory *survey*.

1.2 Justification

The selective collection of solid waste occurs in a diffuse way in a large part of Brazil, due to the anonymous and precarious action of waste pickers, and the lack of knowledge of the importance of collaboration and awareness on the part of the population that generates this waste on a daily basis.

However, even with so many benefits, recycling rates in Brazil are very low, which makes it necessary to encourage selective collection with proper separation of the various materials, both at the time of waste generation and in sorting centers and collectors' cooperatives (GOUVEIA, 2012).

According to Grimberg and Blauth (1998), in order to advance towards environmental, social, economic, political and cultural sustainability in urban waste management, it is essential to evaluate the participation of the population in the various forms of waste collection, separation and disposal that already exist successfully, with the aim of improving the weak points in terms of existing behavior and collaborative thinking.

Since the objective of the PNRS is to share the responsibility of waste generators (manufacturers, distributors, importers, traders and citizens) with the reverse logistics of solid waste, with this in mind, many companies are looking for partnerships with collection cooperatives in order to comply with the law.

However, the importance of this work is due to the lack of studies on the actual level of participation of solid waste generators in the selective collection process, which is why this study aims to provide important data for institutes and other organizations with the intention of inducing reflection on the actions that can be implemented to obtain a response in terms of the population's behavior in participating in the selective collection process.

1.3 Work structure

Below is the structure considered for the work proposed here, which should involve four distinct parts.

In the first part, a bibliographical and documentary survey was carried out on the platforms provided by Anhembi Morumbi University (CAPES, EBSCO Information Services, Google Scholar, Scielo) on the topics of selective collection, collaboration, awareness and solid waste disposal.

In the second part, based on the information gathered, the determining factors were identified for analyzing the level of awareness (such as the price of recyclables, recycling campaigns, laws and regulations on selective collection), and analyzing the level of collaboration of the waste-generating population (such as selective collection points, waste disposal, incentives for selective collection, among others).

In the third part, assertions were created for the analysis of the study aimed at covering the determining factors previously identified, thus generating a *survey* accompanied by an analysis and verification of the results obtained.

The final section presents the conclusions, study limitations, final considerations and suggestions for future work.

2 THEORETICAL FRAMEWORK

2.1 Selective collection

Elucidating the subject of selective waste collection is directly related to the development of content that brings significance to the subject: sustainability.

Over the years, sustainability issues have become a growing concern for humanity (MACEDO, 2016), which highlights the importance of this issue as a central point for reflection on the socio-environmental framework and the dimensions of contemporary societies (JACOBI, 2003).

It encompasses processes related to the vitality of the planet, such as the preservation and conservation of natural resources, "enabling the continuity of life for all species, in parallel with the development of humanity, to meet present and future needs" (CÓRDULA; NASCIMENTO, 2013).

According to Santos (2009), the biggest threat of the 21st century is global climate change, which affects society, the economy and the environment. In view of the effects and vulnerabilities to which the human race is exposed, sustainable attitudes have become decisive factors in relation to environmental deterioration.

In this way, selective collection services present themselves as an alternative with great potential for changing the global scenario, and can act as an agent of transformation in favor of sustainability. But what is selective collection?

Selective collection is the differentiated collection of waste that has previously been separated according to its constitution or composition. In other words, waste with similar characteristics is selected by the generator (which can be a citizen, a company or another institution) and made available for collection separately (BRASIL, [2010]).

This system can be implemented in neighborhoods, homes, schools, offices, shopping centers, among others. However, the municipal cleaning service must be integrated into the project for it to have a more significant impact (CEMPRE 2014). According to Richter (2014), selective collection is an alternative to the problem of waste, providing an opportunity to improve the reuse of solid and organic waste and reduce the volume of waste going to landfill. Thus, investing in the selective collection process brings advantages for environmental costs, these advantages being listed by CEMPRE (2014) as:

- reducing the cost of final waste disposal (landfills or incinerators);
- increasing the useful life of landfills;
- reducing the cost of remediating areas degraded by poor waste disposal (e.g. clandestine dumps);
- environmental education and awareness among the population;
- reduction in general public cleaning costs, considering that the behavior of educated and environmentally aware communities translates into less need for state intervention;
- improving environmental and public health conditions in the municipality.

It is understood that selective collection is the means of separating waste according to its characteristics, thus facilitating its recycling. According to CEMPRE (2014), multi-selective collection (the collection of several types of waste at the same time) faces technical barriers, including the need for its own collection vehicles, physical space where the collected materials will be stored, an increase in collection days, the ability to sell all these materials and the need for educational campaigns.

To mitigate these barriers, greater investments are needed. The PNRS proposes separating dry waste from waste, the former consisting of metals, paper, cardboard, tetra pak and other types of plastic.

With a clearer view of the waste cycle in (conventional) household collection and selective collection, sorting centers are the point in this process where selective collection cooperatives are located. It is within the cooperative that the material is sorted according to its characteristics, and various types of material are found from homes, stadiums, supermarkets, offices and even the door-to-door collection carried out by the municipality.

For Silva (2005), selective waste collection and recycling are indispensable solutions, as they reduce the volume of waste for final disposal in landfills and incinerators. On the other hand, CEMPRE (2016) also highlights the difficulties of conventional collection, which is still cheaper. To put this information into figures, the cost of selective collection (despite the reduction in recent years) is 4.1 times higher than conventional collection. Figure 2.1 shows the selective collection process according to Sebastiao (2011).

Figure 2.1 - The sorting process for selective waste collection

Municipal Solid Waste (MSW)

Central de Triagem (Cooperativa de colata seletiva)

Recyclable waste

Waste Organic

Rejected

Sale

Composting

Final provision

Compound

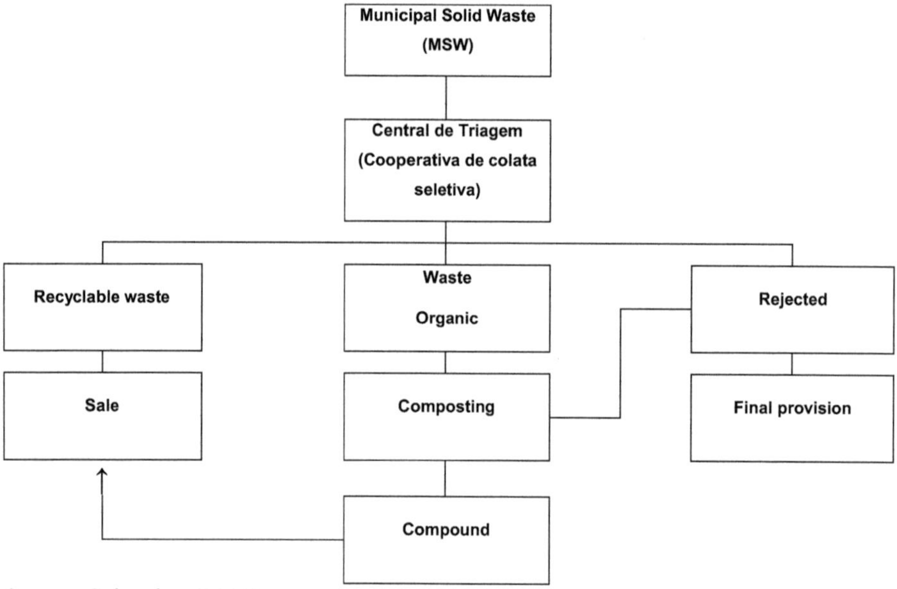

Source: Sebastiao (2011).

Citizens are largely responsible for the generation of solid urban waste, and selective waste collection works as a mechanism for environmental education, making people aware of their consumption and waste of natural resources and the pollution generated by waste discarded unconsciously.

When it receives municipal solid waste, the selective collection cooperative uses its employees to sort the waste in order to separate the recyclable waste (paper, glass, plastics and metals) from the rejects. With the material laid out on conveyor belts, this sorting is done manually, and the final disposal for the waste is landfill and incinerators.

2.2 Solid waste

According to the Brazilian Standard ABNT NBR 10004 - Solid Waste - Classification (2004), solid waste is:

Solid and semi-solid waste resulting from community activities of industrial, domestic, hospital, commercial, agricultural, service and sweeping origin. Included in this definition are sludges from water treatment systems, those generated in pollution control equipment and installations, as well as certain liquids whose particularities make their discharge into the public sewage system or bodies of water unfeasible, or require solutions that are technically and economically unfeasible in view of the best available technology. (NBR 10004/2004)

According to CEMPRE (2016), the gravimetric composition of waste in selective collection is 35% rejects (non-recyclable materials), 11% plastics, 34% paper/paperboard, 6% glass, 2% long life, 3% aluminum, 5% ferrous metals and 4% other, as shown in Figure 2.2.

Figure 2.2 - Gravimetric composition of selective collection

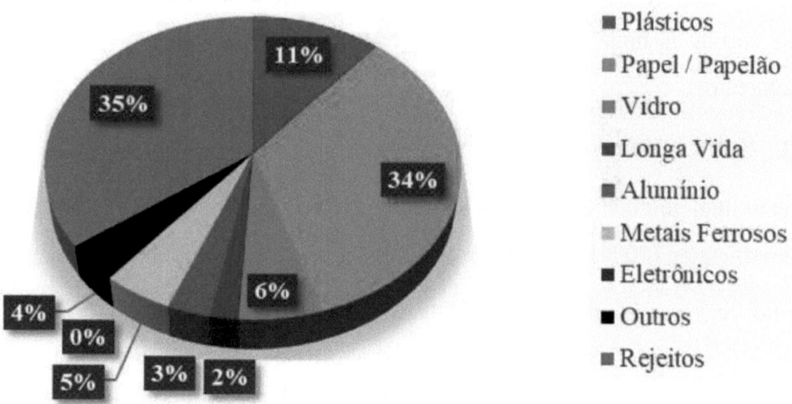

Source: Adapted from Compromisso Empresarial Para Reciclagem (2016).

The generation of Municipal Solid Waste (MSW) has an impact on environmental conservation, hence the need to control and maintain it. In Brazil, 79.9 million tons of MSW were generated in 2015 alone. Of this amount, 72.5 million tons were collected, according to the Brazilian Association of Public Cleaning and Special Waste Companies (ABRELPE, 2015), which represents a collection rate of 90.8% in the country, leaving 7.3 million tons of waste uncollected and therefore with an inappropriate destination.

Studies carried out by Bringhenti (2004) concluded that by investing part of the resources earmarked for the selective collection program in raising awareness among opinion-forming agents and information disseminators - such as reporters and editorial boards of the print media, radio and TV communicators and teachers - these agents could use the indicators to publicize and evaluate selective collection programs. Raising awareness and gaining the support of the media about the importance of using indicators to compare selective collection programs takes time, and the support of this communication channel is fundamental for the consolidation of selective collection in the country.

The author proposes using a portion of the investment from selective collection programs to influence the thinking of people involved in the media about urban solid waste management, so that they can raise public awareness.

According to research by Bringhenti and Gunther (2011) on social participation in selective collection programs, the most important motivating factors for the population are: the environment; the quality of life associated with improved urban cleaning; the reduction of the danger posed by the handling and inadequate disposal of MSW to the staff working in urban cleaning; the organization and adequacy of the operational structure put in place to support the program; the existence of ongoing publicity, mobilization and information actions.

The authors consider that there are certain very important factors that motivate the general population. These must be taken into account and publicized in order to raise awareness among individuals who generate MSW and get them to cooperate in selective collection programs.

2.3 Collaboration

The term collaboration is defined as a process in which people and entities share information, resources and responsibilities together, planning and evaluating a program of activities to achieve a common goal, surrounded by rules and governance to build indicators and monitor their management, which is indicated as a collaborative entity or a collaborative effort (BORGES, 2011).

In addition, it is important to recognize that collaboration strategies are being adopted in various segments, with the aim of innovating and adapting to the rapid changes in the environments in which they operate.

Identifying and characterizing the value system of the community and its members is fundamental to improving collaboration, because the decision-making process in a collaborative community is naturally influenced by both the common value of the community system and the individual value of each part of the system. For this reason, identifying and characterizing these value systems is an important issue because, as the partners have different value systems, they may have different perceptions of the outcome, which in some cases can lead to non-collaborative behaviour and intra-organizational conflicts (CAMARINHA-MATOS; MACEDO, 2010).

In Science and Technology Research and Development, Silva (2007) comments that successful collaboration can evolve into cooperation, because this adds functions to the partners, without being limited to the sectoral differences between them, for example.

The collaboration of the population in the selective collection process is very important for the implementation of each city or country's waste management model. Waste recycling must necessarily come from selective collection. It is therefore necessary for citizens to take an active part in sorting waste in their own homes.

Within the theme of collaboration, it is necessary for all citizens to have the culture of contributing to the selective collection process, sorting waste at home and depositing it in the appropriate containers, or separating it on the day indicated by the council of each city, with the ultimate aim of improving the quality of said separation and achieving greater and better recycling of the different materials. This translates into environmental benefits resulting from savings in natural resources, raw materials, energy and atmospheric emissions (CAMARINHA- MATOS; ABREU, 2007).

To achieve this culture of collaboration, it is essential to know that the population has to be sensitized through constant environmental education campaigns, so that society as a whole makes selective collection a sustainable habit over time (BORGES, 2011).

This is a process that takes time, but it is possible to get people to take action through activities that teach and explain that society's responsibility in terms of consumption is to separate the waste generated in the domestic environment with the highest possible quality and then deposit it in the correct collection points (BORGES, 2011). Carrying out these actions properly conditions the entire subsequent waste management process. Therefore, this action will be constituted as a behavior for the environment and will have a more significant impact as more people join together to carry out these same actions.

2.4 Awareness

Awareness is defined as the action and effect of creating awareness in people about a problem or phenomenon that is considered important, and from then on, habits and attitudes can be changed so that they can adjust to the new reality. Awareness-raising activities alone cannot guide sustainable transformation. This process begins with sharing information with people about the topic you want to raise awareness of, and in this way raising awareness in order to achieve a change in the population's behavior and attitude. Only after this knowledge has been acquired and expanded in the communities is it possible to have an organization willing to create actions that can counteract the existing problem (LORENZ, 2009).

Nowadays, with the emergence of environmental issues and the association of consumerism with waste, new trends have emerged in society. Sustainability, as well as awareness, has become part of people's thinking and consumers tend to no longer submit to the rules of the market.

A survey on conscious consumption carried out by FECOMERCIO in Rio de Janeiro, in partnership with IPSOS (2012), revealed that in 2011 certain unsustainable practices by Brazilians were still quite common, with 23.1% of the population believing that the responsibility for global warming due to bad environmental habits does not lie with them. It can be seen then that the main objective, according to this study, is to acquire skills, create a sense of responsibility and stimulate individual and collective action for the separation of waste for selective collection and that it depends not only on an individual attitude, but also on conditions provided by governments and companies to be realized.

According to Maletzke's model (1963 apud SANCHEZ, ABACHE, DOMINGUEZ, 2013), a person who knows about recycling, its causes and consequences, can raise awareness and develop in social structures where the message is shared around it and thus make people aware of their surroundings, which is the ultimate impact of the campaign. This is why the public must be involved in the stage that is their responsibility, which is to raise awareness of the need to separate waste for selective collection, a fundamental procedure for obtaining a recycled product that is as good as the original. If separation is carried out efficiently at home, at work and in the community, there will be benefits such as cleanliness and hygiene in cities and an improved quality of life for the population, lower production costs for industries that reuse recyclable material, savings on electricity, wealth generation, the creation of new jobs, the strengthening of community organizations, etc.

We are currently living in a time of great transformations: social, economic, political, environmental, among others, and if there isn't an environmental attitude and awareness, repairing the damage caused to the environment and avoiding new ecological disasters, the continuity and quality of life will be compromised. This would be the biggest mistake humanity could make against itself (VIEIRA, 2004).

In a campaign launched in Madrid, Spain, in February 2016, under the name "SEPARATE, RECYCLE, LOVE Madrid", organized by the Madrid City Council, in partnership with the ECOEMBES organization, different associated actions were designed to create awareness among the population through campaigns based on interpersonal communication:

- Information for citizens in the street: one of the main actions of the environmental educators was to walk through the streets, squares and public spaces of the city's 21 districts, explaining to neighbors how to properly separate their waste and resolve any doubts they may have about this;
- Personalized advice on housing, offering real alternatives to the needs of neighbors to facilitate the separation of their waste at home;
- Communicating directly with doormen, inspectors and representatives of the housing communities, they are very important agents in the management of waste in our city. The environmental educators have visited neighboring farms, providing information and resolving doubts, as well as helping with any difficulties they may encounter in their work related to waste;
- Taxi drivers: they have joined the campaign as key allies, providing their

customers with the "Questions when sorting your waste?" guide. inside their vehicles, so that travelers can find out more during their journeys;

- Hairdressers: the campaign for them was aimed at resolving the doubts that clients and professionals may have about how to properly segregate waste;
- Training and workshops for different groups, associations, Adult Education Centers (CEPAs), Employment Agency courses, etc;
- Theaters: Informing passers-by of the benefits of recycling and the importance of separating waste;
- Presence at events, celebrations and actions in the neighborhood with stimulating and supportive interventions on the subject of waste, as well as on all the platforms and social networks that were at the Campaign's disposal.

3 RESEARCH METHODOLOGY

Below is the methodological structure considered for this work in terms of nature, approach, objectives, methods, data collection techniques and data analysis techniques, summarized in Table 1.

TABLE 1 - Classification of scientific methodology.

Classification according to the nature of the research	**Classification according to the approach used**	**Classification according to the objectives of the work**	**Classification according to the method used**	**Classification by data collection technique**	**Classification by data analysis technique**
Applied.	Mixed.	Exploratory.	*Survey.*	Bibliographical research; Documentary research; Questionnaire; Interview.	Descriptive statistical analysis.

In terms of nature, this research is characterized as applied, as Marconi and Lakatos (2010) identify applied, empirical or field research as that used to obtain information and knowledge about: a certain problem for which an answer is sought; proof of a hypothesis; discovery of new phenomena or relationships between them. Since this research is a means of obtaining information and knowledge about the level of collaboration and awareness of the population regarding selective waste collection, it can be classified as applied.

In terms of research approach, a mixed approach is one in which the researcher "persuasively and rigorously collects and analyzes both qualitative and quantitative data (based on the research questions)", according to Creswell and Clark (2013). As this work includes the collection and analysis of both qualitative and quantitative data, it can be said that the approach of this research is characterized as mixed.

In terms of objectives, exploratory research aims to bring the researcher closer to the problem in order to form hypotheses or make it more explicit, with flexible planning. This type of research can involve: bibliographical research; interviews; analysis of examples that can stimulate understanding (GIL, 2002). Therefore, as this research has the characteristics mentioned above, seeking to bring researchers closer to the problem of raising awareness about selective waste collection and its aspects, its classification in terms of the objectives of the work is revealed as exploratory.

The research method is classified as a *survey*, collecting data through questionnaires applied to samples, supported by sampling techniques and analyzed through statistical inference (NAKANO, 2012). The exploratory *survey*, according to Forza (2002), involves collecting information from individuals about themselves or the organizations to which they belong. As this work uses data collected by means of a questionnaire applied to a sample of the population and analyzed statistically, its research method is reaffirmed as a *survey*.

As far as data collection techniques are concerned, documentary research was carried out, obtaining data by analyzing documents on the problem studied. Bibliographic research was also used, involving consultations of scientific publications relating to the field of knowledge of interest to the research (MARCONI, 2010).

Also in terms of data collection, a questionnaire with multiple choice answers was made available on a social network. The sample size was estimated with a maximum error of 5% and 80% reliability, requiring a minimum of 164 respondents.

Descriptive statistics and correlation were used to analyze the data. The aspects considered for analyzing the level of understanding and cooperation of individual waste generators in the selective collection process are listed in Table 2.

TABLE 2 - Study categories for the selective collection process.

CategQria/fatQΓ		Description
Awareness	Cipation for packaging and re ci ve products	This category allows us to assess waste generators' preference for recyclable products/packaging. Is there a preference for recyclable products and packaging at the time of purchase?
	Price of recyclables	This category allows us to assess whether the prices of recyclable products/packaging iπfluencian∩ waste generators at the time of purchase. Do the prices of recyclable products and packaging influence waste generators at the time of purchase?
	Campaigns to re ci c l age ni	This category allows us to assess whether waste generators are influenced by separate collection and recycling campaigns. Have waste generators had contact with/influence from collection and recycling campaigns recently?
	Laws and regulations on selective collection and re ci cl age ni	This category allows us to assess whether waste generators are aware of or have heard of selective collection standards and laws. Do waste generators know about regulations and laws for the disposal and collection of solid waste?
Collaboration	Selective collection points	This category allows you to assess the availability of selective collection points. Are selective collection points available near the home/work of waste generators?
	Waste disposal	This category allows us to assess whether waste generators separate recyclable and organic waste at the time of disposal. Do you separate your recyclable and organic waste at the time of disposal?
	Separate collection notices	This category makes it possible to assess whether waste generators are influenced by separate collection and recycling notices. Do waste generators obey the separate collection notices available at work and in public places?
	Encouraging selective collection	This category makes it possible to assess whether waste generators are encouraged in their social environment to separate their waste at the time of disposal. Would waste generators choose to separate their waste for financial compensation?

4 Discussion and analysis of data

The data used for the analysis was collected from the responses to the questionnaire, which included 21 questions designed by the authors for the respondents. These questions were divided into two categories: awareness and collaboration of waste generators. The answers were given according to a Likert scale: (1) totally disagree, (2) disagree more than agree, (3) neither agree nor disagree, (4) agree more than disagree and (5) totally agree. Questions were also asked to characterize the respondents: gender, age, level of education and occupation. 170 responses were obtained for this study.

The first question characterized 68% of respondents as female and 32% as male, as shown in Figure 4.1.

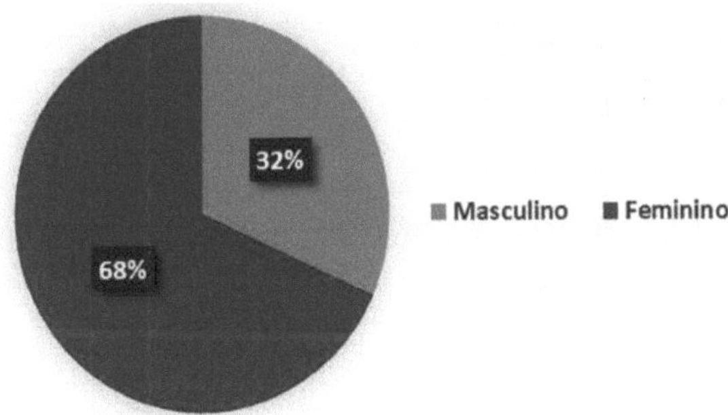

Figure 4.1 - Gender.

The second question analyzed the age of the respondents: 22% aged between 10 and 20, 47% aged between 21 and 30, 16% aged between 31 and 40, 7% aged between 41 and 50 and 8% of respondents aged over 50, as shown in Figure 4.2. There is a predominantly young profile between the ages of 21 and 30.

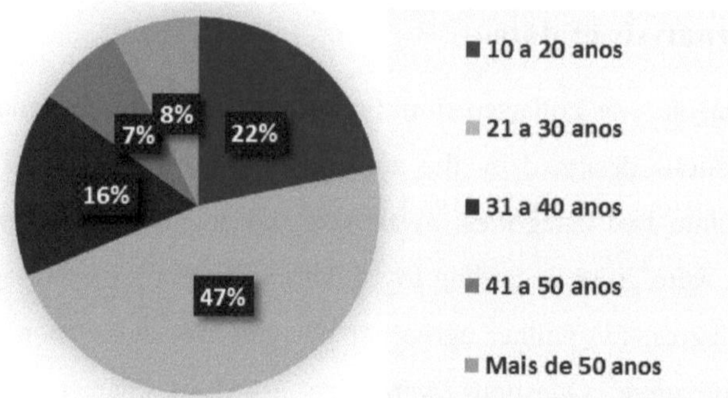

Figure 4.2 - Age.

In the third question, illustrated in Figure 4.3, the respondents' level of education was identified: 1% had incomplete primary education, 4% had completed primary education, 1% had incomplete secondary education, 13% had completed secondary education, 47% had incomplete higher education and 34% had completed higher education.

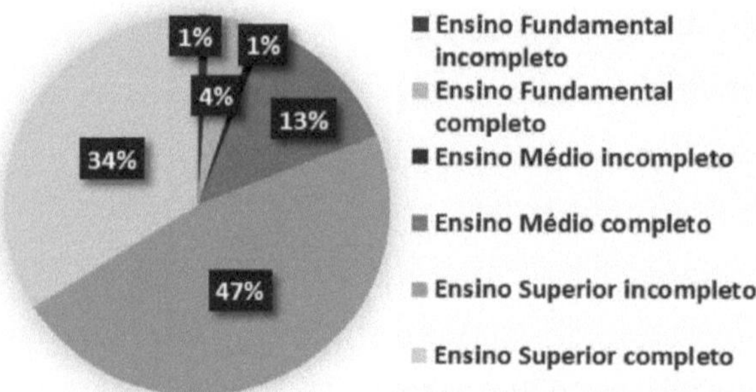

Figure 4.3 - Level of education.

In the fourth question, illustrated in Figure 4.4, the respondents' level of occupation was determined, where 32% only study, 32% only work, 30% work and study and 6% neither work nor study. In terms of occupation, the profile of the respondents is diverse.

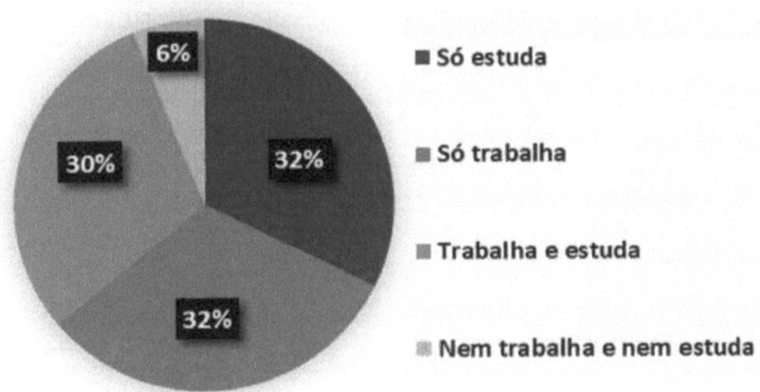

Figure 4.4 - Occupation.

Once the respondents had been characterized, the category concerning the level of awareness of waste generators was checked. Initially, each respondent's personal knowledge of the amount of waste they generate in a week, a month or a year was analyzed. The results showed that 54% totally disagree, 14% disagree more than agree, 19% neither agree nor disagree, 9% agree more than disagree, and 4% totally agree. These results show that only a small proportion of respondents are aware of the amount of waste they generate.

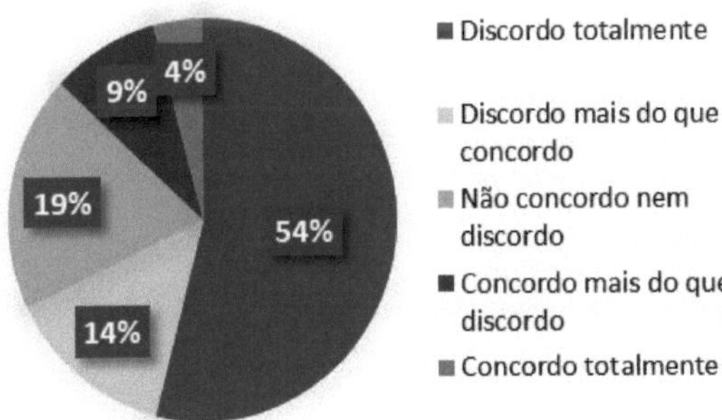

Figura 4.5 - Knowledge of the amount of waste generated.

As part of the same awareness-raising question, respondents were asked how desirable it was to have selective collection points everywhere. As shown in Figure 4.6 below, 1% totally disagree, 2% disagree more than agree, 8% neither agree nor disagree, 11% agree more than disagree, and 78% totally agree. The vast majority of respondents agree that it is desirable to have separate collection points everywhere.

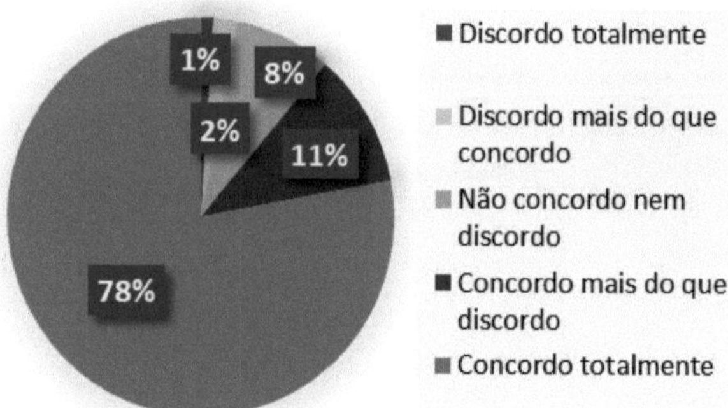

Figura 4.6 - The desire for separate collection points everywhere.

Next, we analyzed whether respondents prefer recyclable packaging and products when making purchases. The results showed that 9% totally disagreed, 8% disagreed more than they agreed, 43% neither agreed nor disagreed, 21% agreed more than they disagreed and 19% totally agreed, as shown in Figure 4.7. Given the results, 43% of respondents are indifferent about their preference for recyclable packaging and products. Only 19% of respondents actually have such a preference.

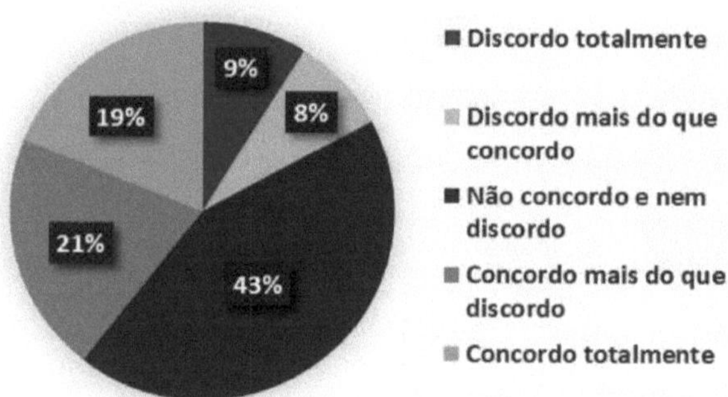

Figura 4.7 - Preference for recyclables.

The next question asked whether recyclable products and packaging are no longer purchased when their prices are higher than competing products and packaging. The following responses were obtained: 11% totally disagree, 8% disagree more than agree, 22% neither agree nor disagree, 22% agree more than disagree and 37% totally agree. Given these results, the majority agree that they would not buy recyclable products and packaging if they were priced higher than competing products and packaging, as shown in Figure 4.8.

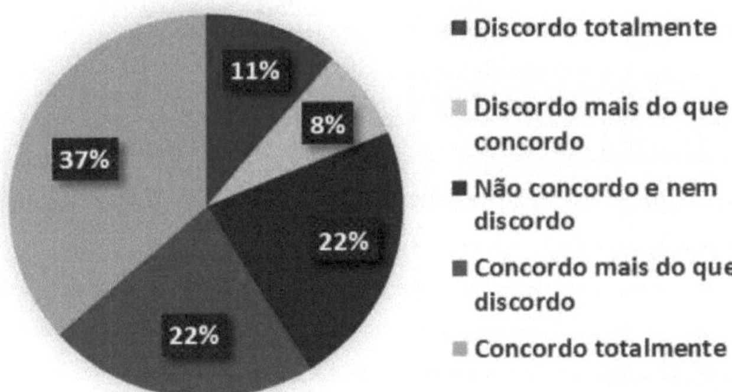

Figura 4.8 - Influence of the prices of recyclable products and packaging.

Still on the subject of awareness, the next question looked at the influence of awareness campaigns for selective collection and recycling. Respondents were asked whether collection and recycling campaigns influence the separation of waste. The results are illustrated in Figure 4.9: 4% totally disagree, 5% disagree more than agree, 23% neither agree nor disagree, 27% agree more than disagree and 41% totally agree. It can therefore be said that collection and recycling campaigns need to be stepped up in order to achieve better results in selective collection and recycling processes.

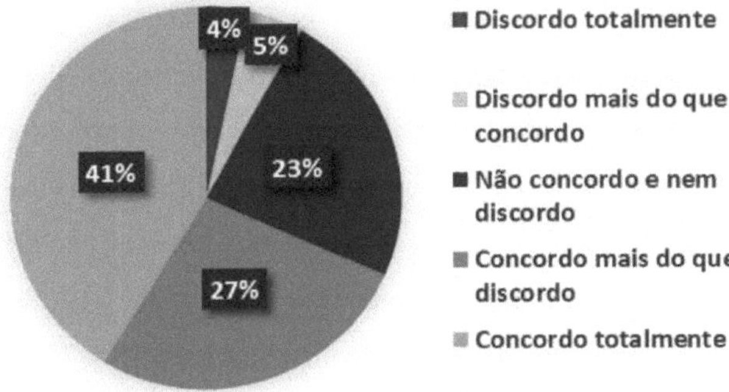

Figura 4.9 - Influence of collection and recycling campaigns.

Given the relevance of the collection and recycling campaigns demonstrated in the previous question, the next question looked at how often these awareness campaigns for selective collection and recycling had been noticed by the respondents who generate waste in recent times. From the answers obtained, it was found that: 24% of respondents said they had never noticed a selective collection and recycling campaign, 24% of respondents said they had had contact in the last 3 months, 16% in the last 6 months and 21% had not had contact with this type of campaign for over a year, as illustrated in Figure 4.10. The results point to the need to step up selective collection and recycling campaigns.

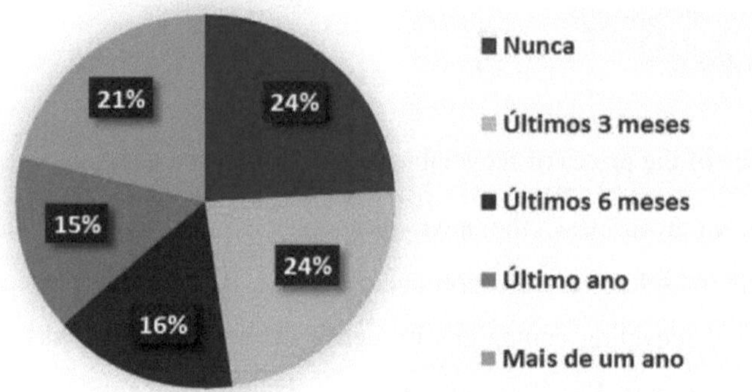

Figure 4.10 - Frequency of collection and recycling campaigns.

To conclude the verification of the level of awareness of the respondents who generate waste, it was checked whether selective collection can be boosted by laws and regulations on this subject. We checked whether the respondents were aware of laws and regulations on selective collection and recycling and the results were: 45% were totally unaware, 15% were partially unaware, 15% were neither aware nor unaware, 13% were partially aware and 12% were totally aware. The results therefore show that respondents' knowledge of laws and regulations on selective collection and recycling is low. Only a small proportion of respondents claim to be fully aware of the laws and regulations on the subject in question, as illustrated in Figure 4.11.

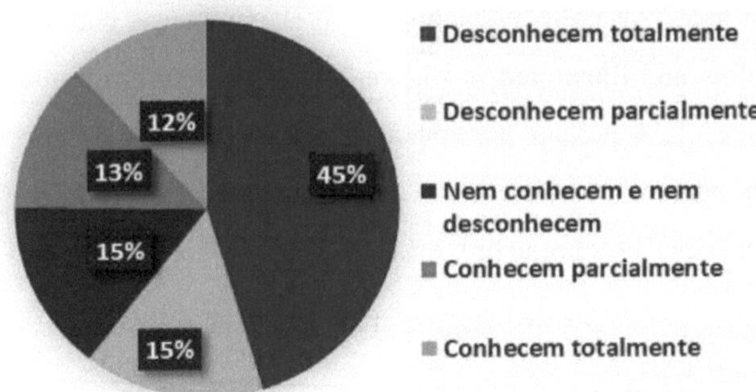

Figura 4.11 - Knowledge of laws and regulations on selective collection and recycling.

Next up was the category of waste generators' collaboration in selective collection and recycling. To check this category, the following points were considered: selective collection points, waste disposal, notices of selective collection and encouraging colleagues to selective collection.

The respondents were first asked about the availability of collection points for selective waste collection near their homes. The results obtained indicated that: 34% totally disagree, 18% disagree more than agree, 21% neither agree nor disagree, 14% agree more than disagree, and 13% totally agree. It can be seen that the largest proportion of respondents (34%) totally disagreed with the availability of collection points for selective waste collection near their home.

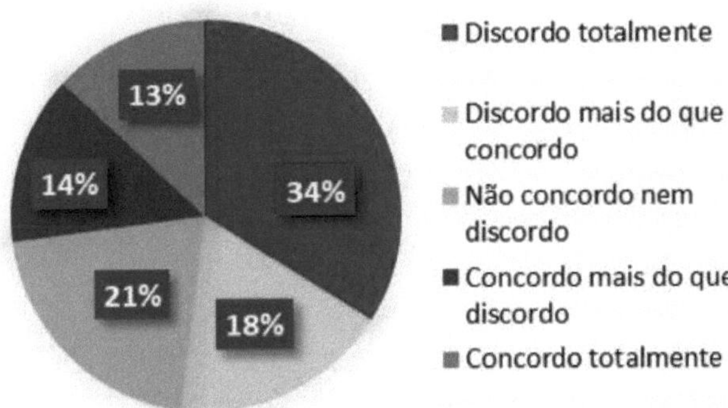

Figura 4.12 - Availability of selective collection points near the home.

Next, the availability of selective collection points at their work was checked. The results are presented below and illustrated in Figure 4.13: 39% totally disagree, 8% disagree more than agree, 25% neither agree nor disagree, 12% agree more than disagree and 16% totally agree. According to the answers obtained, 39% do not have selective collection points near their home or work.

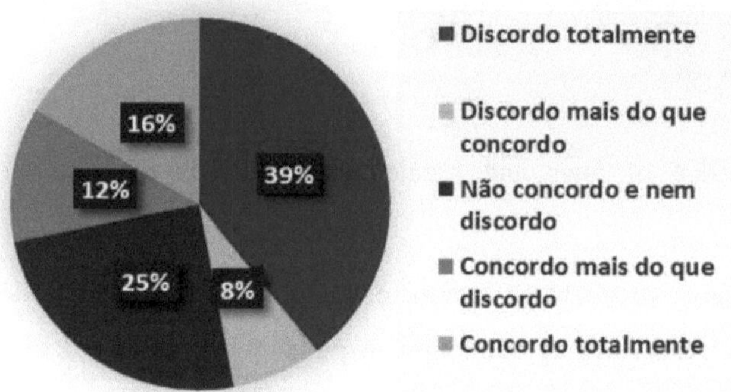

Figura 4.13 - Availability of collection points close to work.

Still in the context of collaboration, the next question asked respondents whether they knew of any selective collection cooperatives near their home or workplace. The results are illustrated in Figure 4.14 and indicate that: 52% totally disagree, 11% disagree more than agree, 14% neither agree nor disagree, 11% agree more than disagree, and 12% totally agree. This shows that the majority of respondents (52%) totally disagree that they know of a selective collection cooperative near their work or home.

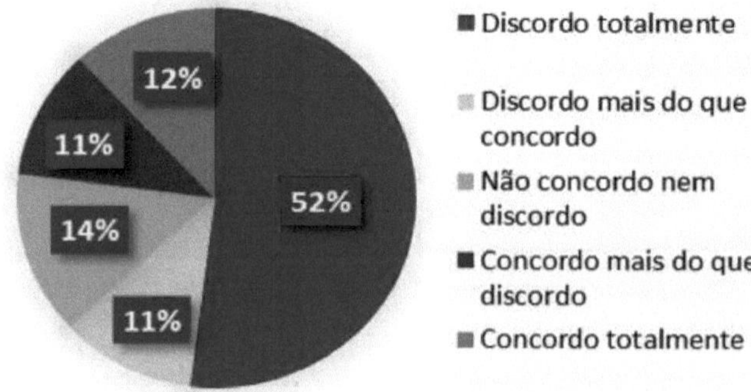

Figura 4.14 - Knowledge of nearby selective collection cooperatives.

The next question, illustrated in Figure 4.15, looked at the respondents' waste separation practices. Respondent waste generators were asked how they assessed their waste separation practices at the time of disposal, and the results were: 38% considered it bad, 46% considered it fair, 12% considered it good and 4% considered it excellent.

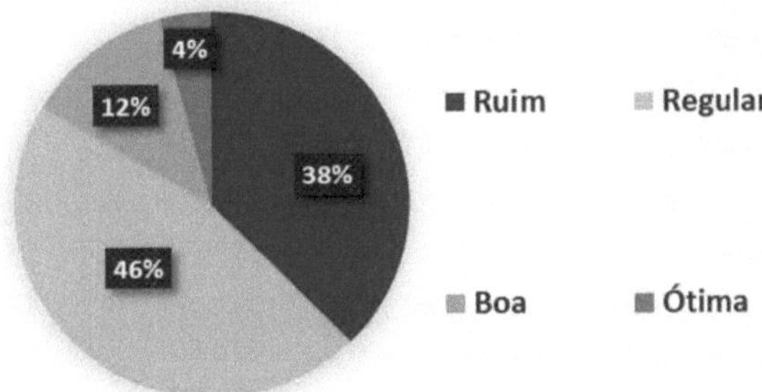

Figure 4.15 - Waste separation practice.

Next, the respondents who generate waste were asked whether they obey the selective collection notices when they are available. The results were: 1% totally disagreed, 2% disagreed more than agreed, 14% neither agreed nor disagreed, 16% agreed more than disagreed and 67% totally agreed. The majority of respondents said they obeyed the separate collection notices when they were available, as illustrated in Figure 4.16.

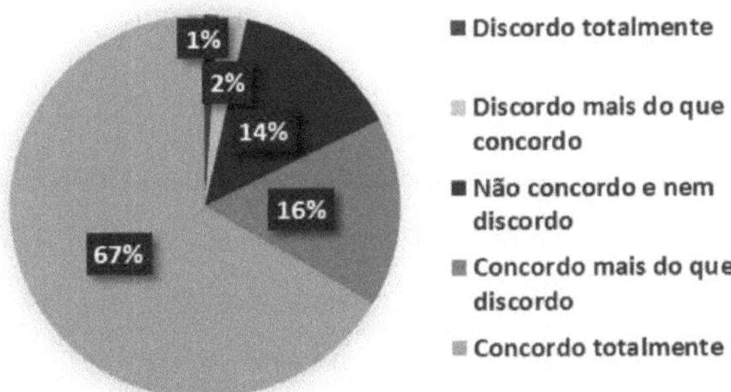

Figura 4.16 - Selective collection notices.

Finally, the question was asked whether the separation of waste could be motivated by financial compensation. The results were: 7% totally disagree, 3% disagree more than agree, 15% neither agree nor disagree, 21% agree more than disagree and 54% totally agree, as shown in Figure 4.17. With financial compensation, the majority of respondents would choose to separate waste at the time of disposal.

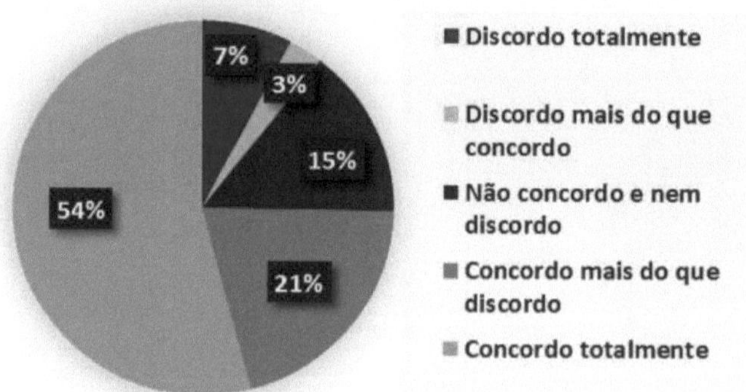

Figura 4.17 - Encouraging selective collection.

The majority of respondents who generate waste are young people aged between 21 and 30 and have completed or incomplete higher education. Only 6% do not work or study.

With regard to opting for recyclable products and packaging, 43% of respondents were indifferent, but 37% of these respondents confirmed in another question that the price difference is a determining factor in whether or not they opt for these products. It is believed that when recyclable products and packaging are more expensive, they are no longer purchased by respondents.

Separate collection and recycling campaigns were highly relevant in motivating waste separation practices, according to 41% of respondents. Despite this, 24% of respondents said they had never seen this type of campaign and 21% said they had not seen this type of campaign for more than a year. Since the campaign is a motivating factor for selective collection, there is a need to step up selective collection and recycling campaigns, given that 45% of respondents also said they were unaware of the laws and regulations applicable to the subject.

Collection points are essential for sorting waste in selective collection. Despite this, 39% of respondents said that there were no collection points available at their work or near their home. This demonstrates the need to expand the availability of selective collection points. It may be that this lack of collection points corroborates respondents' poor (38%) and regular (46%) performance in terms of waste separation practices. Even so, 67% of respondents said they obeyed the selective collection notices when they were available.

The majority of waste generators (54%) stated that they would improve waste separation if there was financial compensation for doing so.

In short, a summary of the information collected in the awareness category can be seen in Table 3, and in the collaboration category in Table 4.

TABLE 3 - Data collection in the awareness category.

Approach	Answers				
The prices of packaged products recyclables influence you at the time of buy?	I disagree completely 24,1%	I disagree more with that agree 14.1%	I disagree nor disagree 28,8%	I agree more than that disagree 19,4%	I agree completely 13,5%
Recyclable packaging products are no longer be purchased when they are more expensive?	11,2%	7,6%	22,4%	22,4%	36,5%
Do you have a preference for recyclable products/packaging?	8,8%	8,2%	43,5%	20,6%	18,8%
Recycling campaigns influence waste separation behavior?	3,5%	4,7%	23,5%	27,1%	41,2%
Do you know of any laws or regulations for collecting selective and solid waste disposal?	45,3%	15,3%	14,7%	12,9%	11,8%
Do you know how much solid waste you generate in a week, a month or a year?	54,1%	14,1%	18,8%	9,4%	3,5%
Do you think that a disposal area for selective collection should be available in every places?	1,2%	2,4%	7,6%	10,6%	78,2%
Do you care about selective waste collection? waste disposal?	0,6%	4,1%	18,2%	17,6%	59,4%
In your opinion, companies are currently concerned with selective waste collection waste?	16,5%	17,6%	35,3%	18,2%	12,4%
When was the last time you noticed a selective waste collection campaign in your social environment?	Never 24,1%	Last 3 months 23,5%	Last 6 months 15,9%	Last year 15,3%	More than one year 21,2%

TABLE 4 - Data collection in the collaboration category.

Approach	Answers				
Is there a service available in your city? selective collection?	I disagree completely	I disagree more with that agree	I disagree nor disagree	I agree more than that disagree	I agree completely
	13,5%	11.2%	27.1%	27,6%	20,6%
Are there separate collection points near your home?	34,3%	18,3%	20,7%	13,6%	13,0%
Are selective collection points available at your workplace?	39,4%	7,6%	24,7%	11,8%	16,5%
Do you separate your recyclable and organic waste when disposing of it?	24,7%	16,5%	14,1%	12,9%	31,8%
Do you respect and obey waste collection notices at work and in public places?	1,2%	2,4%	14,1%	15,9%	66,5%
Is there any benefit in your social environment to practicing waste separation? time of disposal?	34,7%	19,4%	18,8%	10,6%	16,5%
Do you encourage the people around you to sort their waste? time for disposal?	15,9%	12,9%	22,9%	21,8%	26,5%
Would you sort your waste if there was financial compensation for doing so?	7,6%	2,9%	14,7%	20,6%	54,1%
Are there separate collection notices at your work?	37,1%	14,7%	17,1%	10,0%	21,2%
Do you know of any selective collection cooperatives near your home or work?	51,8%	11,2%	14,1%	11,2%	11,8%
How do you rate public participation? in selective collection ?	Great	Good	Regular	Bad	
	4,1%	12,4%	45,9%	37,6%	

5 Conclusions

In this study, various perceptions and behaviors of the interviewees in relation to collaboration in the selective collection process were identified, as well as the motivating factors for such participation.

The conclusion is that the important factors in raising awareness and cooperation among individuals are: lower prices for recyclable products and packaging, more frequent selective collection and recycling campaigns and financial compensation for separating waste.

Based on descriptive statistics, it was possible to verify that the respondents' collaboration in waste separation and selective collection has performed poorly due to the lack of availability of collection points, lack of knowledge about laws and regulations on selective collection and recycling, little practice in waste separation, and campaigns to improve their awareness.

The limitations of this research are mainly the time it took to collect the data. It is hoped that the results obtained can contribute to directing campaigns and solutions for waste separation and selective collection. It is suggested that further research be carried out into the motivational factors for waste separation and selective collection identified in this study.

6 REFERENCES

BRAZILIAN ASSOCIATION OF PUBLIC CLEANING AND Special Waste COMPANIES - ABRELPE. **Panorama of solid waste in Brazil 2015**. Available at: < http://www.abrelpe.org.br/Panorama/panorama2015.pdf>. Accessed on: April 22, 2017.

BORGES, Wagner Luis. **An analysis of the implementation of CPFR in retail companies.** 2011. 115 f. Dissertation (Master's Degree) - Production Engineering Course, University of Sao Paulo, Sao Carlos, 2011. Available at: <http://www.teses.usp.br/teses/disponiveis/18/18156/tde-02052012-114021/publico/Wagner_Luis_Borges.pdf>. Accessed on: May 29, 2017.

BRAZIL. Law No. 12.305, of August 2, 2010. **Institutes the national solid waste policy**. Official gazette of the union, Brasilia, 2010. Available at: <fld.com.br/catadores/pdf/politica_residuos_solidos.pdf>. Accessed on: April 16, 2017.

BRAZIL. MINISTRY OF THE ENVIRONMENT. **Selective Collection.** [2010]. Available at: <http://www.mma.gov.br/cidades-sustentaveis/residuos-solidos/catadores-de-materiais-reciclaveis/reciclagem-e-reaproveitamento>. Accessed on: October 18, 2017.

BRINGHETI, Jacqueline. **Selective collection of solid urban waste: operational aspects and public participation.** 2004. 235 f. Thesis (Doctorate) - Public Health Course, Environmental Health, Faculty of Public Health, University of Sao Paulo, Sao Paulo, 2004. Available at: <https://www.researchgate.net/profile/Wanda_Maria_Guenther/publication/266471400_COLETA_SELETIVA_DE_RESIDUOS_SOLIDOS_URBANOS_ASPECTOS_OPERACIONAIS_E_DA_PARTICIPACAO_DA_POPULACAO_ORIENTADOR_PROF_a_DR_a/links/56699bd008aea0892c49ad0c.pdf>. Accessed on: 05 May 2017.

BRINGHENTI, Jacqueline R.; GÜNTHER, Wanda M. Risso. Social participation in selective urban solid waste collection programs. **Engenharia Sanitària e Ambiental,** [s.l.], v. 16, n. 4, p.421-430, dez. 2011. FapUNIFESP (SciELO). http://dx.doi.org/10.1590/s1413-41522011000400014. Available at: <http://www.scielo.br/pdf/esa/v16n4/a14v16n4.pdf>. Accessed on: 05 May 2017.

CAMARINHA-MATOS, L. M.; ABREU, A.. Performance indicators for collaborative networks based on collaboration benefits. **Production Planning & Control,** [s.l.], v. 18, n. 7, p.592-609, 7 sep. 2007. Informa UK Limited.

http://dx.doi.org/10.1080/09537280701546880. Available at:

<http://www.tandfonline.eom/doi/full/1 0.1080/0953728070 15 46880?scroll=top&needAccess=true>. Accessed on: September 22, 2017.

CAMARINHA-MATOS, L. M.; MACEDO, P. **A conceptual model of value systems in collaborativenetworks**. Journal Intell Manuf, 2010. 287

299p. Available at:

<https://pdfs.semanticscholar.org/2c9c/da67b3370806e2b6211890019567388493cc.pdf> Accessed on: 01 Nov. 2017.

CEMPRE (Sao Paulo). **Radiographing selective collection**, 2016. Available at: <http://cempre.org.br/ciclosoft/id/8>. Accessed on: March 22, 2017.

CEMPRE (Sao Paulo). **A portrait of 20 years of selective waste collection in the country**, 2016.

Available at: <http://cempre.org.br/cempre-informa/id/7/um-retrato-de-20-anos-da-coleta-seletiva-no-pais>. Accessed on: March 22, 2017.

CICLOSOFT (Sao Paulo). **Radiographing Selective Collection.** 2016. Available at: <http://cempre.org.br/ciclosoft/id/8>. Accessed on: March 22, 2017.

CÓRDULA, Eduardo Beltrao de Lucena; NASCIMENTO, Glòria Cristina Cornélio do. THE HERMENEUTICS OF ENVIRONMENTAL EDUCATION AND THE PARADOX OF SUSTAINABILITY. **Revista Eletrônica em Gestâo, Educaçâo e Tecnologia Ambiental,** [s.l.], v. 8, n. 8, p.1573-1580, 9 jan. 2013. Universidad Federal de Santa Maria. http://dx.doi.org/10.5902/223611706670. Available at: <https://periodicos.ufsm.br/reget/article/download/6670/pdf>. Accessed on: 14 Apr. 2017.

CRESWELL, John W.; CLARK, Vicki L. Plano. **Mixed Methods Research.** 2. ed. Porto Alegre: Penso, 2013. 288 p.

FORZA, Cipriano. Survey research in operations management: a process-based perspective. **International Journal Of Operations & Production Management,** [s.l.], v. 22, n. 2, p.152-194, feb. 2002. Emerald.

http://dx.doi.org/10.1108/01443570210414310. Available at: <https://edisciplinas.usp.br/pluginfile.php/1107460/mod_resource/content/1/Forza Survey.pdf>. Accessed on: Nov. 20, 2017.

GIL, Antonio Carlos. **How to prepare research projects.** 4. ed. Sao Paulo: Atlas, 2002. 176 p. Available at:

<https://professores.faccat.br/moodle/pluginfile.php/13410/mod_resource/content/1/como_elaborar_projeto_de_pesquisa_-_antonio_carlos_gil.pdf>. Accessed on: 09 Oct. 2017.

GOUVEIA, Nelson. Urban solid waste: socio-environmental impacts and the prospect of sustainable management with social inclusion. **Ciência & Saùde Coletiva,** [s.l.], v. 17, n. 6, p.1503-1510, jun. 2012. FapUNIFESP (SciELO). http://dx.doi.org/10.1590/s1413-81232012000600014. Available at: <http://www.scielo.br/pdf/csc/v17n6/v17n6a14.pdf>. Accessed on: October 30, 2017.

GRIMBERG, Elisabeth; BLAUTH, Patricia. **Selective Collection: Recycling Materials, Recycling Values.** 31. ed. Sao Paulo: Pòlis, 1998. 100 p. Available at: <http://www.polis.org.br/uploads/984/984.pdf>. Accessed on: October 29, 2017.

(IBGE), Brazilian Institute of Geography and Statistics. **National Survey of Basic Sanitation - PNSB 1989**. Rio de Janeiro, 1994. Available at: <https://biblioteca.ibge.gov.br/biblioteca-catalogo.html?view=detalhes&id=284395>. Accessed on: November 1, 2017.

IGLESIAS, Francisco. **Industrial revolution**. 3. ed. Sao Paulo: Brasiliense, 1982. 114 p.

JACOBI, Pedro et al. **Environmental education, citizenship and sustainability**. Cadernos de pesquisa, v. 118, n. 3, p. 189-205, 2003. Available at: <www.scielo.br/pdf/cp/n118/16834.pdf>. Accessed on: April 8, 2017.

LORENZ, Viviane Esther. **Art and environmental awareness: a reflection on continuing teacher training, based on Basil Bernstein.** 2009. 205 f. Thesis (Doctorate) - Human Sciences Course, Federal University of Rio Grande do Sul, Porto Alegre, 2009. Available at: <http://www.lume.ufrgs.br/bitstream/handle/10183/21392/000737105.pdf7sequence=1>. Accessed on: May 21, 2017.

MACEDO, Maria Alexandra; RAMOS, Maria da Conceiçao Pereira. **Education**

Environmental and Urban Solid Waste: The Way Forward

Sustainable. Eduser - Revista de Educaçao, [S.l.], v. 7, n. 2, dec. 2016. Available at: <https://www.eduser.ipb.pt/index.php/eduser/article/view/65>. Accessed on: May 22, 2017.

MADRID. MADRID CITY COUNCIL. **Separate, Recycle, Want to Madrid.** 2016. Available at:

<http://www.madrid.es/portales/munimadrid/es/Inicio/Medio-ambiente/Residuos-y-limpieza-urbana/Campana-para-el-fomento-de-la-separacion-de-residuos?vgnextfmt=default&vgnextoid=c3d57ec79a3c1510VgnVCM2000000c205a0aRCRD&vgnextchannel=f81379ed268fe410VgnVCM1000000b205a0aRCRD&WT;.ac=Campana_de_separacion_de_residuos_2016>. Accessed on: May 15, 2017.

MARCONI, Marina de Andrade. LAKATOS, Eva Maria. **Fundamentals of scientific methodology**. 7 ed. Sao Paulo, Atlas, 2010. 320 p.

NAKANO, Davi. **Research methods adopted in production engineering and operations management**. 2 ed. Sao Paulo, Elsevier, 2012. 280 p.

Brazilian Standard ABNT NBR 10004 - Solid Waste - Classification. 2004.

Available at: <http://www.unaerp.br/documentos/2234-abnt-nbr-10004/file>.

Accessed on: October 18, 2017.

PORTILHO, Fâtima. **Environmental sustainability, consumption and citizenship.** Sao Paulo: Cortez Editora, 2005. 256 p.

RICHARD KOLLODGE (United States). United Nations Population Fund (Ed.). **World Population Situation Report 2011**: People and possibilities in a world **of** 7 billion. New York: Unfpa, 2011. 126 p.

Available at:

<http://www.portaldoenvelhecimento.com/uploads/images/janeiro2012/Pessoas-e-possibilities-in-a-world-of-7-billion .pdf>. Accessed on: September 25, 2017.

RICHTER, Leonice Terezinha. **THE IMPORTANCE OF AWARENESS AND SELECTIVE COLLECTION IN THE MUNICIPALITY OF PALMITOS - SC.** 2014. 78 f Monograph (Specialization) - Environmental Management in Municipalities Course, Federal Technological University of Paranà, Medianeira, 2014. Available at: <http://repositorio.roca.utfpr.edu.br/jspui/handle/1/4537>. Accessed on: Aug. 15, 2017.

RIO DE JANEIRO. Superintendence of Economics and Research. Rio+20 Representation Commission, 2nd Public Hearing (Org.). **Contributions from the Fecomércio-RJ System.** 2012. Available at: <https://www.slideshare.net/alerj/comisso-de- representao-rio20-2-audincia-pblica-contribuçies-do-sistema-fecomrciorj-22-de-maro- de-2012>. Accessed on: May 15, 2017.

SANCHEZ, Francis; ABACHE, Kelly; DOMiNGUEZ, Jonathan. **Communication Model:** Gerald Maletzke. 2013. 12 f. Research Paper - Higher Education Course, Universidad Católica Cecilio Acosta, Maracaibo, 2013. Available at: <https://es.slideshare.net/abachekelly/modelo-de-comunicacin-maletzke-teor>. Accessed on: October 29, 2017.

SANTOS, Filipe Duarte. The Environmental Challenges Created by the Great Post-War Acceleration. **Naçâo e Defesa,** Lisboa, v. 4, n. 122, p.61-78, abr. 2009. Available at: <https://comum.rcaap.pt/bitstream/10400.26/3629/1/NeD122_FilipeDuarteSantos.pdf>. Accessed on: April 16, 2017.

SÂO PAULO. Bruno Covas. Secretariat for the Environment (Org.) **Selective collection.** 2013. Available at:

<http://arquivos.ambiente.sp.gov.br/cea/2014/11/coleta-seletiva-em-comunidade-empresa-escola-condominio.pdf&rct=j&sa=U&ved=2ahUKEwiJh56Ehuf ZAhWCIZAKHXIAAicQFjAAegQIBxAB&q=coleta+seletiva+secretaria+do+meio+ambiente&usg=AOvVaw2VlJXIcH1s_D-9o2ibbDIv>. Accessed on: August 12, 2017.

SEBASTIÂO, R. F. **Small municipalities on the north coast of Rio Grande do Sul and the national solid waste policy.** Final paper presented to the environmental engineering course at the federal university of rio grande do sul as part of the requirements for obtaining the title of environmental engineer, 2011.

SILVA, Darly Henriques da. International cooperation in science and technology: opportunities and risks. **Revista Brasileira de Politica Internacional,** [s.l.], v. 50, n. 1, p.5-28, jun. 2007. FapUNIFESP (SciELO). http://dx.doi.org/10.1590/s0034-73292007000100001. Available at: <http://www.scielo.br/pdf/rbpi/v50n1/a01v50n1.pdf>. Accessed on: September 22, 2017.

SILVA, L. T.; DAMASCENA, J. S. **Cooperativism and solid waste recycling: an approach based on the principles of solidarity economy**.

Unitrabalho, 2005. Available at: <http://www.unitrabalho.uem.br/artigos/pdf8.pdf>. Accessed on: May 6, 2017.

VICTOR BICCA (São Paulo). Cempre Compromisso Empresarial Para Reciclagem (Ed.). **Guide to selective waste collection.** 2014. Available at:

<https://www.google.com.br/url?url=http://cempre.org.br/download.php?arq=b18xOXE2ZTQxcnFpbTgxdGc2cnA2cWczdmV0YS5wZGY=&rct=j&sa=U&ved=2ahUKEwiB2PSjiufZAhUEGpAKHdycAWYQFjAAegQIBxAB&q=Guia+da+coleta+seletiva+de+lixo&usg=AOvVaw3gLhfyYIM5-GZXiJhmgqjy>. Accessed on: September 20, 2017.

VIEIRA, Lênia Ribeiro de Souza. **Environmental awareness at the Saint-Gobain plumbing factory: an environmental education methodology for industry.** 2003. 161 f. Thesis (Doctorate) - Engineering Course, University of Sao Paulo, Sao Carlos, 2004. Available at: <http://www.teses.usp.br/teses/disponiveis/18/18139/tde-11082016- 153947/publico/Tese_Vieira_LeniaRS.pdf>. Accessed on: May 18, 2017.

VIEIRA, Raquel Fonseca. **Solid waste management in the context of municipal environmental management in Varginha, MG: challenges and potential.** 2008. 169 f. Dissertation (Master's Degree) - Administration Course, Federal University of Lavras, Lavras, 2008. Available at:

<http://repositorio.ufla.br/bitstream/1/2579/1/DISSERTAÇÂO_Gestào de residuos sólidos no contexto de gestao ambiental municipal em Varginha, MG.pdf>. Accessed on: 01 Nov. 2017.

Printed by Books on Demand GmbH, Norderstedt / Germany